SHAPES
Symmetry

Activity Workbook

DRAW & COLOR

Ages 4+

USE AT YOUR OWN RISK

⚠ WARNING: CHOKING HAZARD

 All rights reserved. No part of this publication may be reproduced, distributed, or transmitted, in any form or by any means, including photocopying, recording, or other electronic or mechanical methods, without prior written permission of the publisher, expect in the case of brief quotations embodied in critical reviews and certain other noncommercial uses permitted by copyright law.

THIS BOOK BELONGS TO:

_____

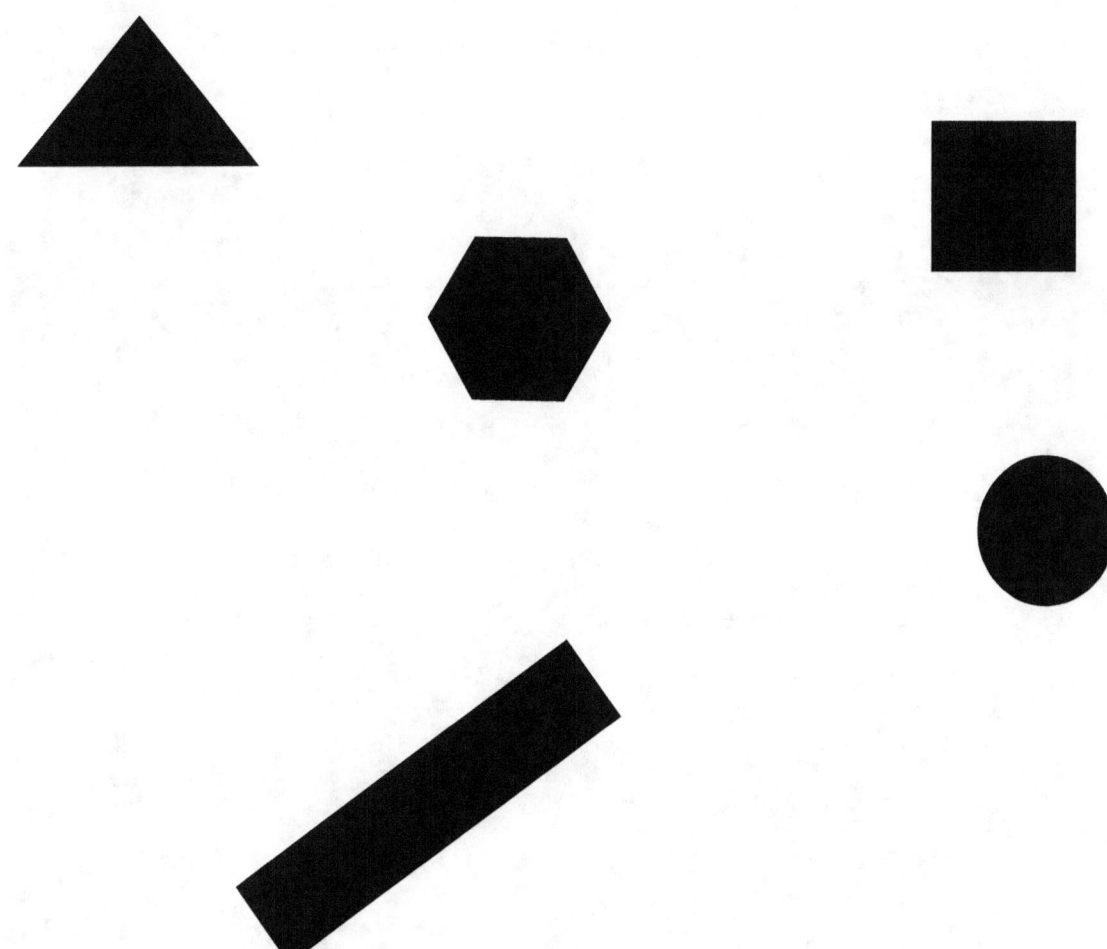

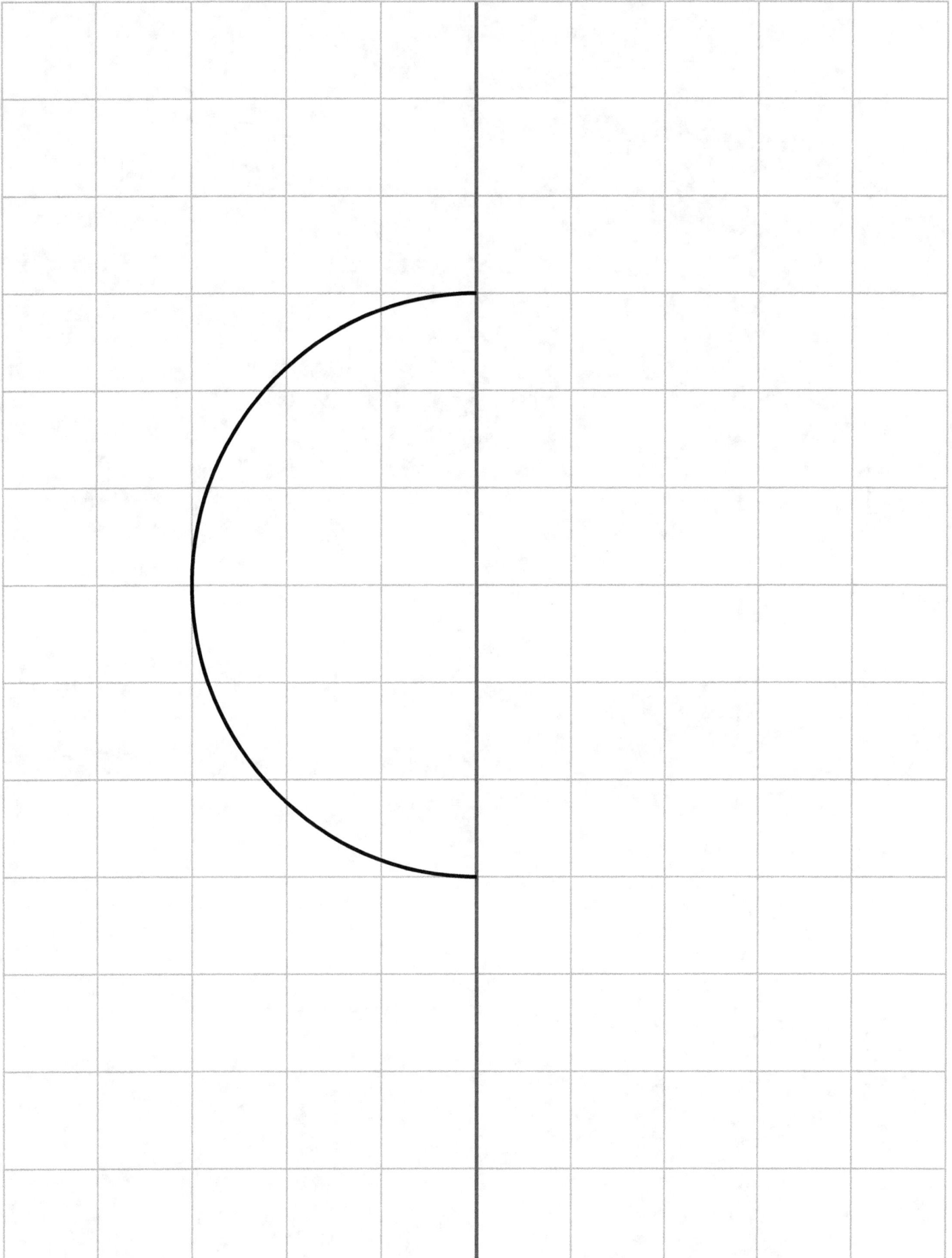

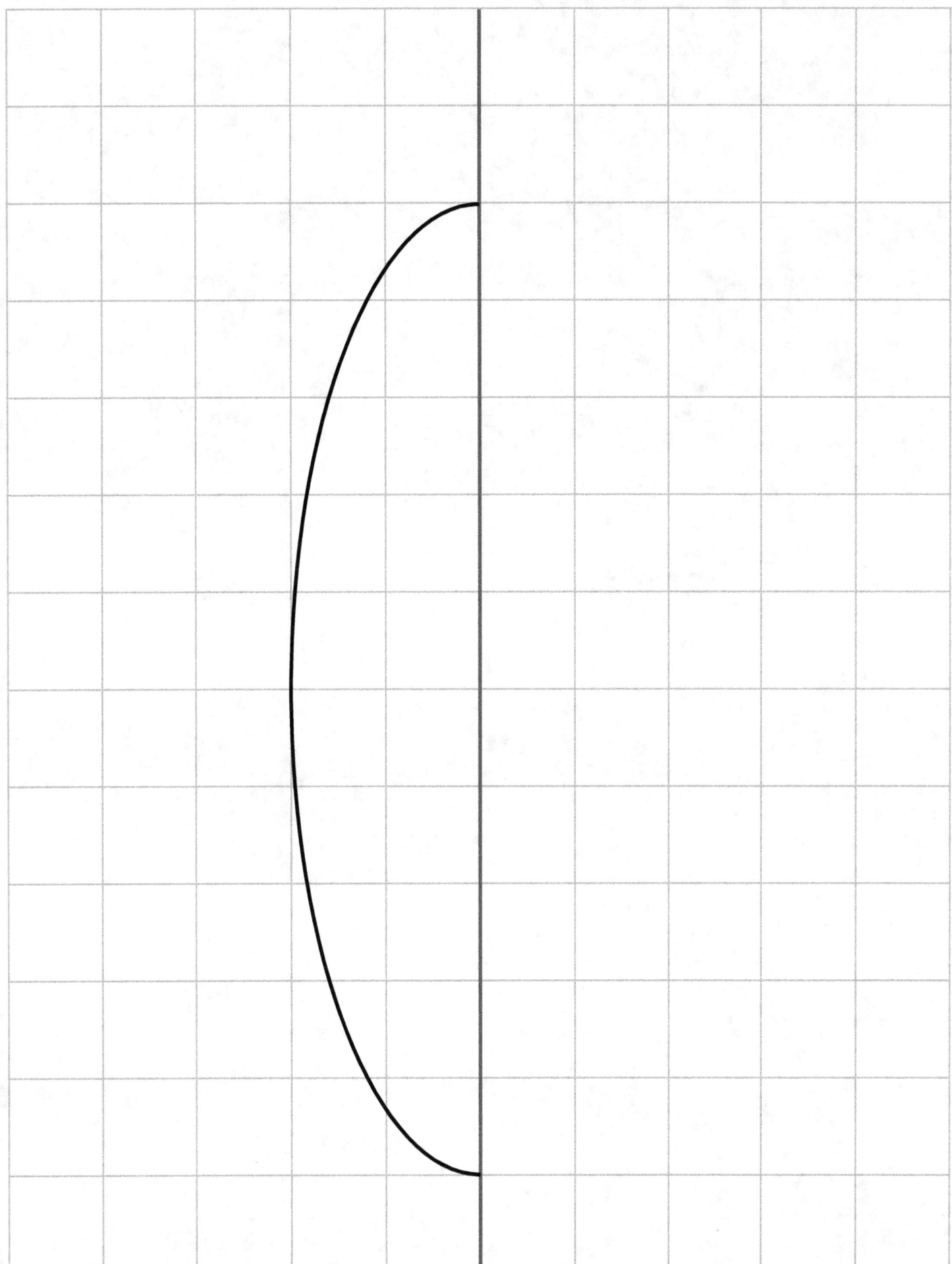

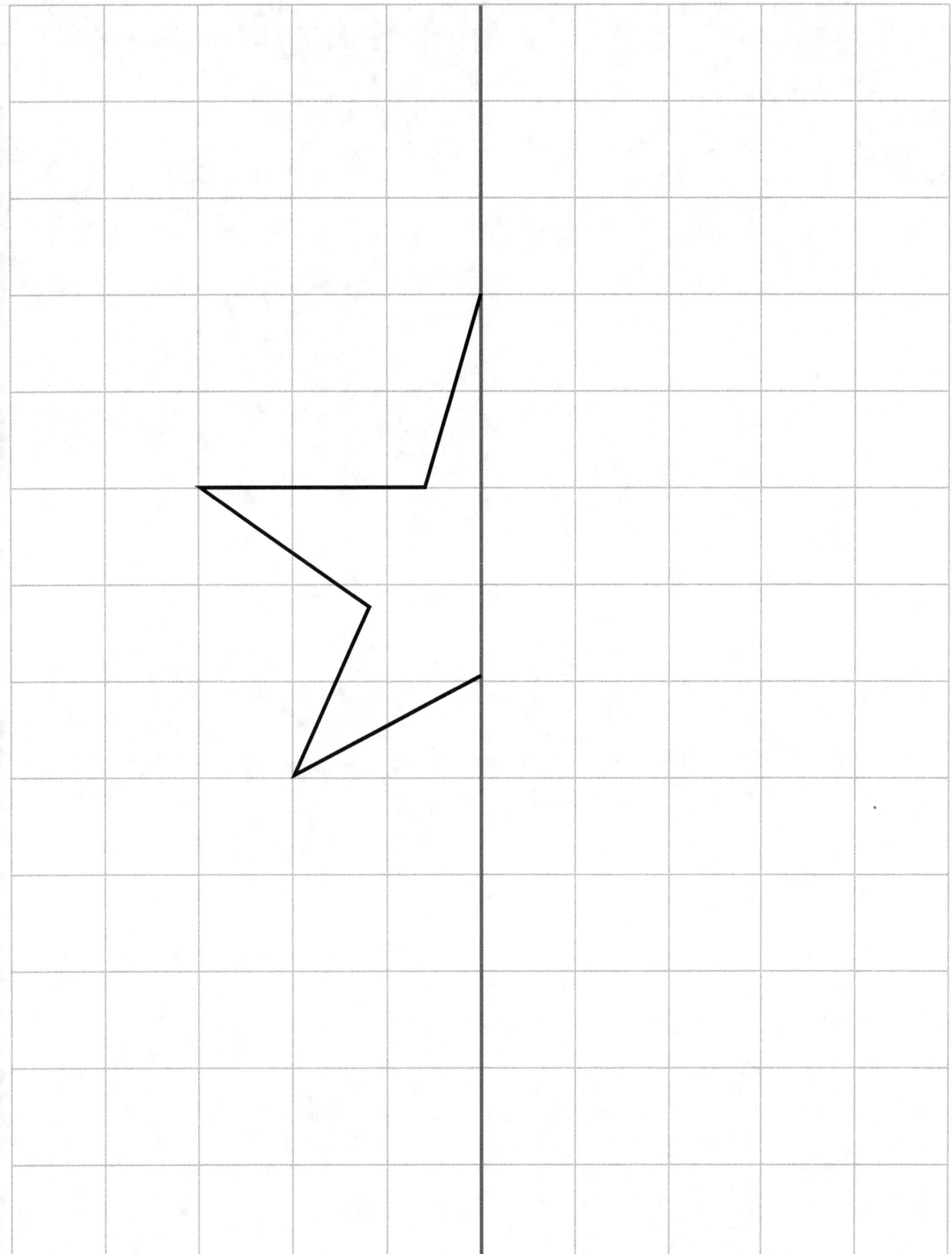

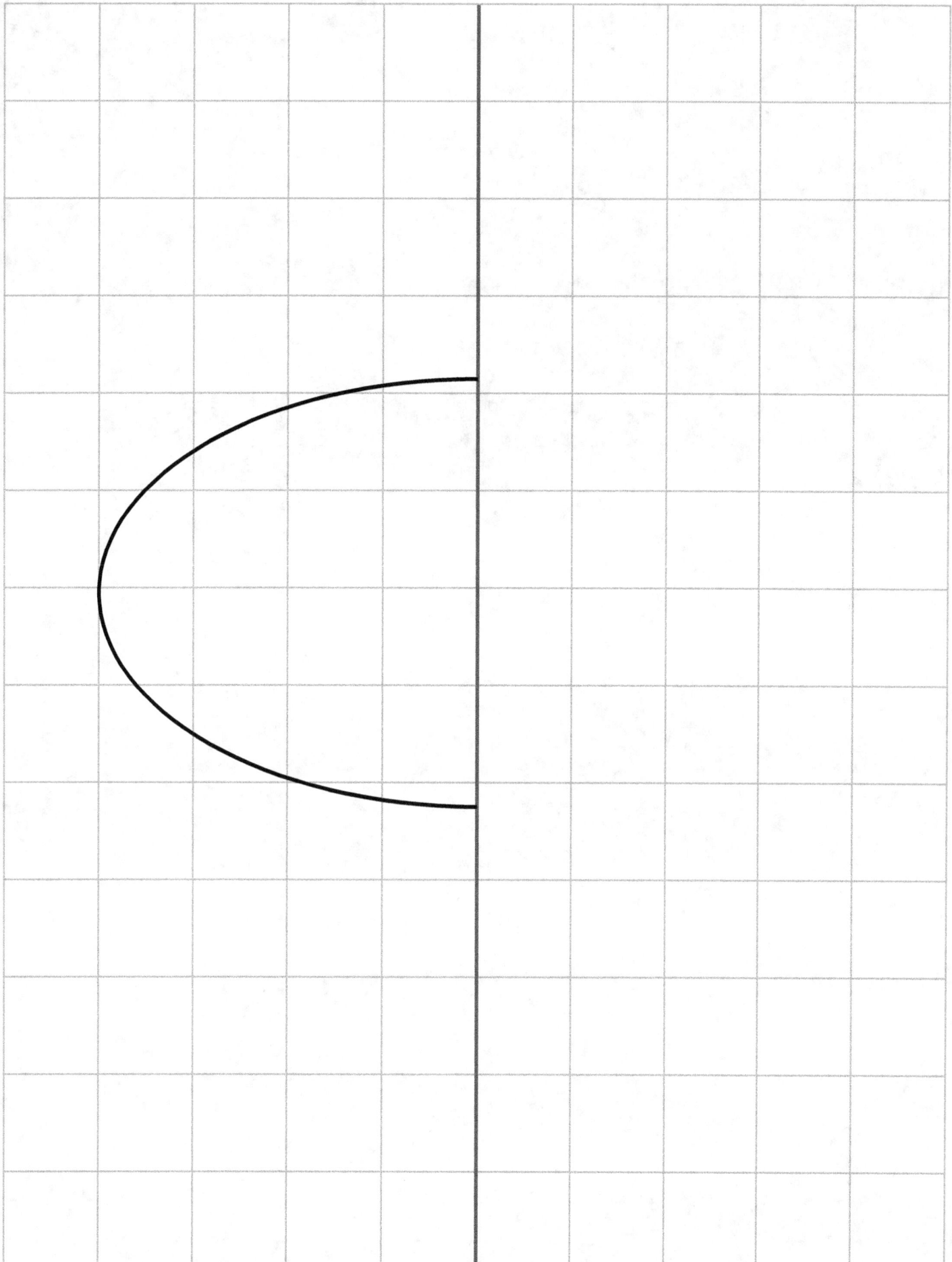

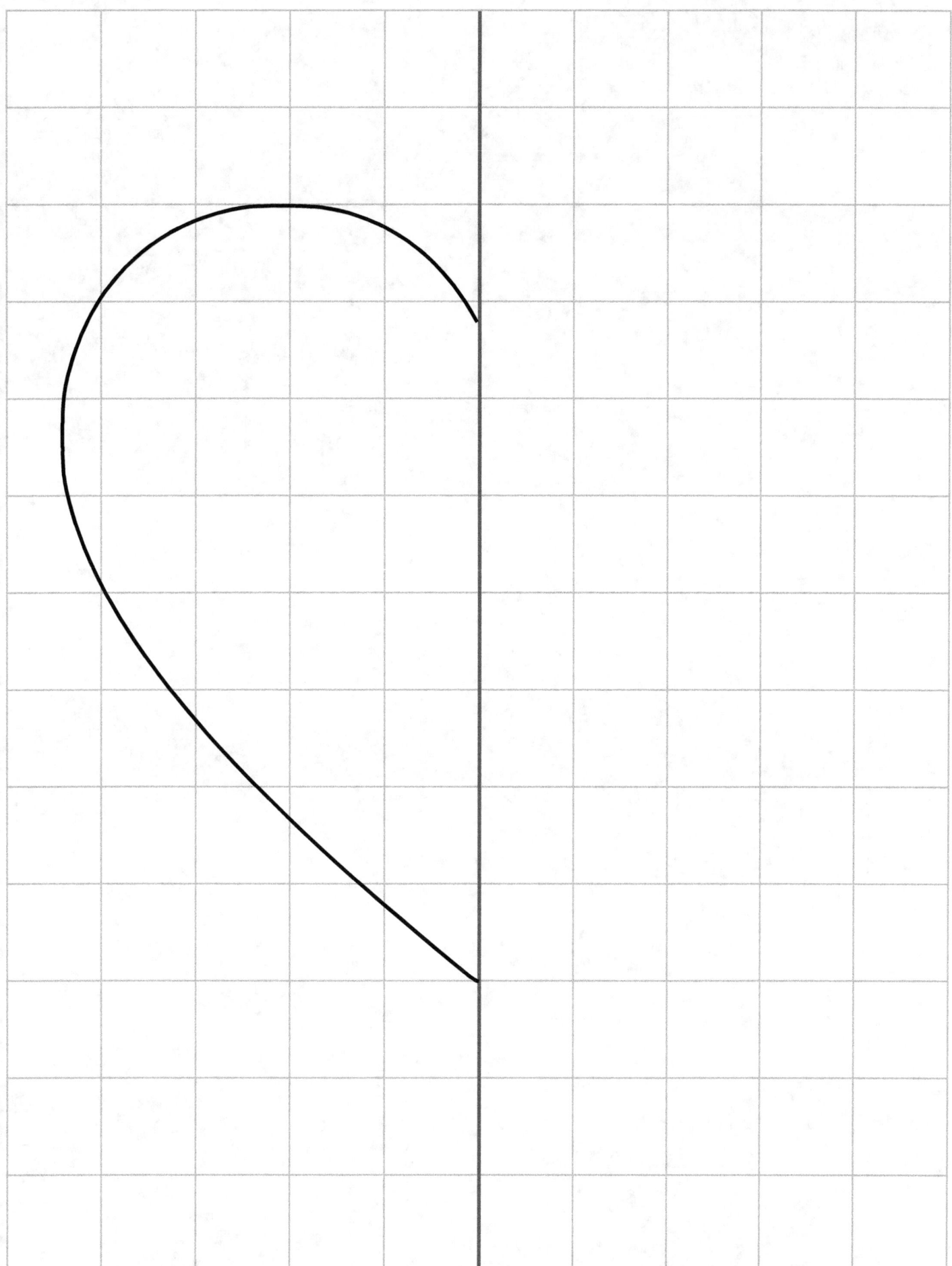

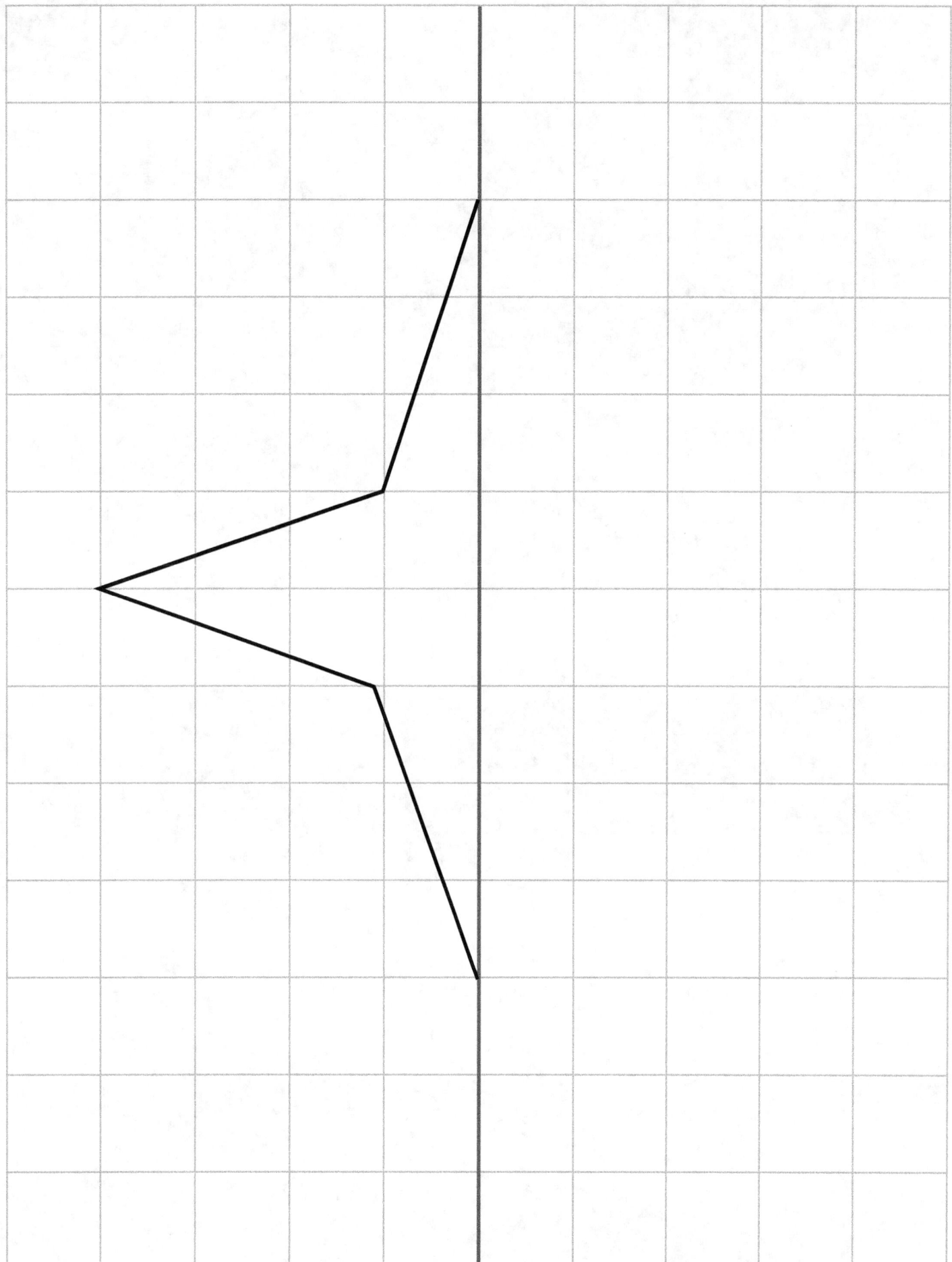

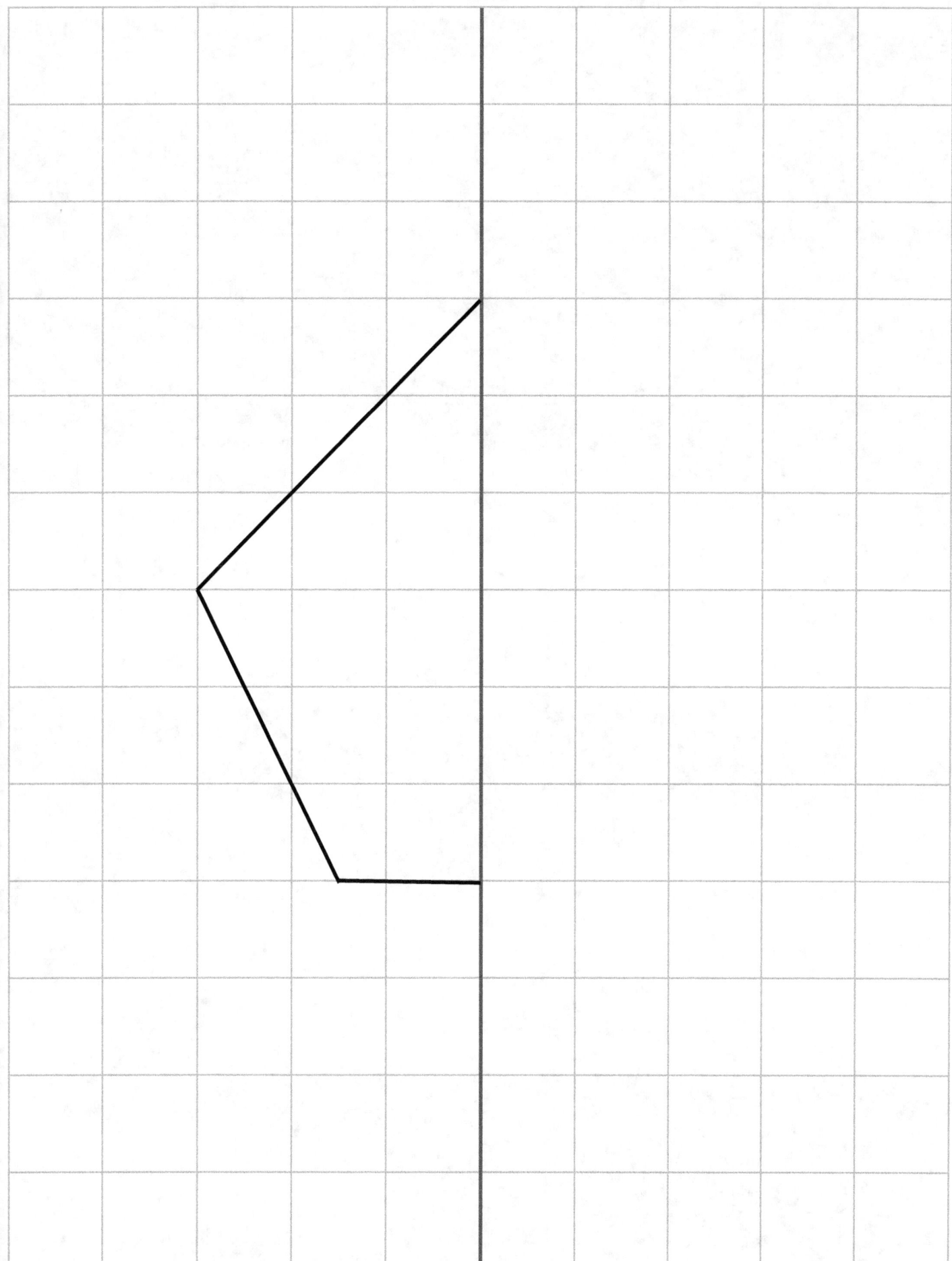

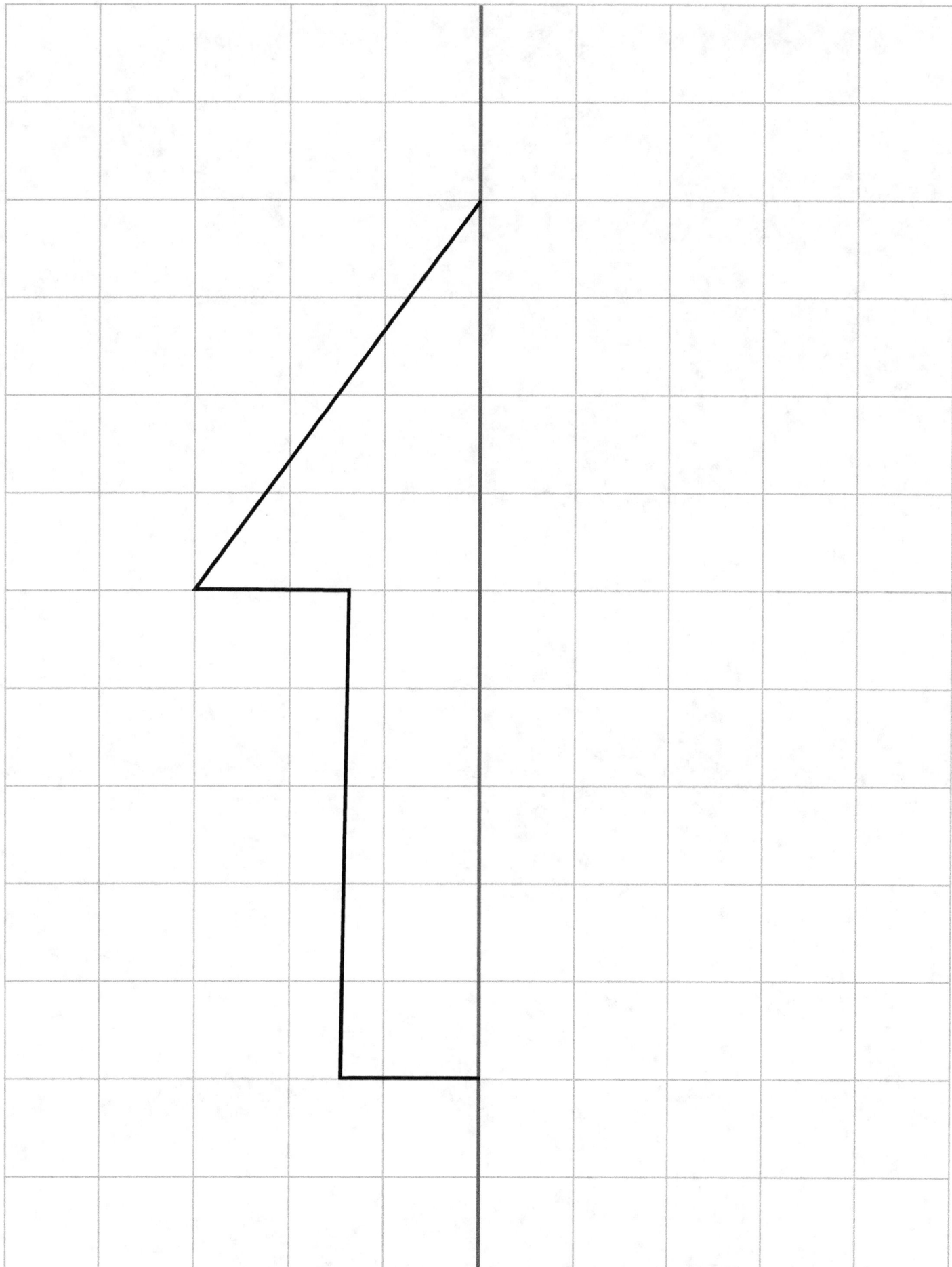

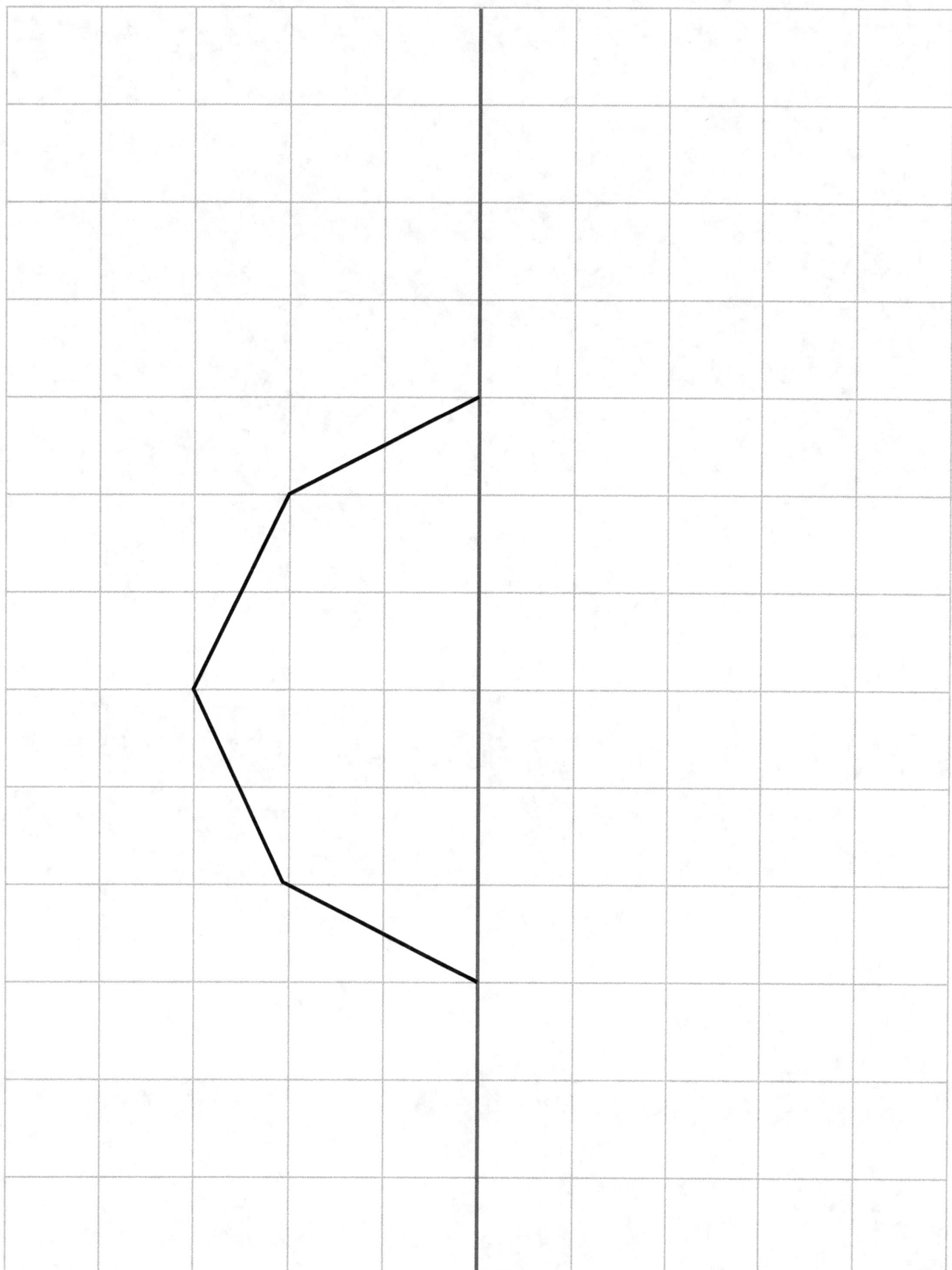

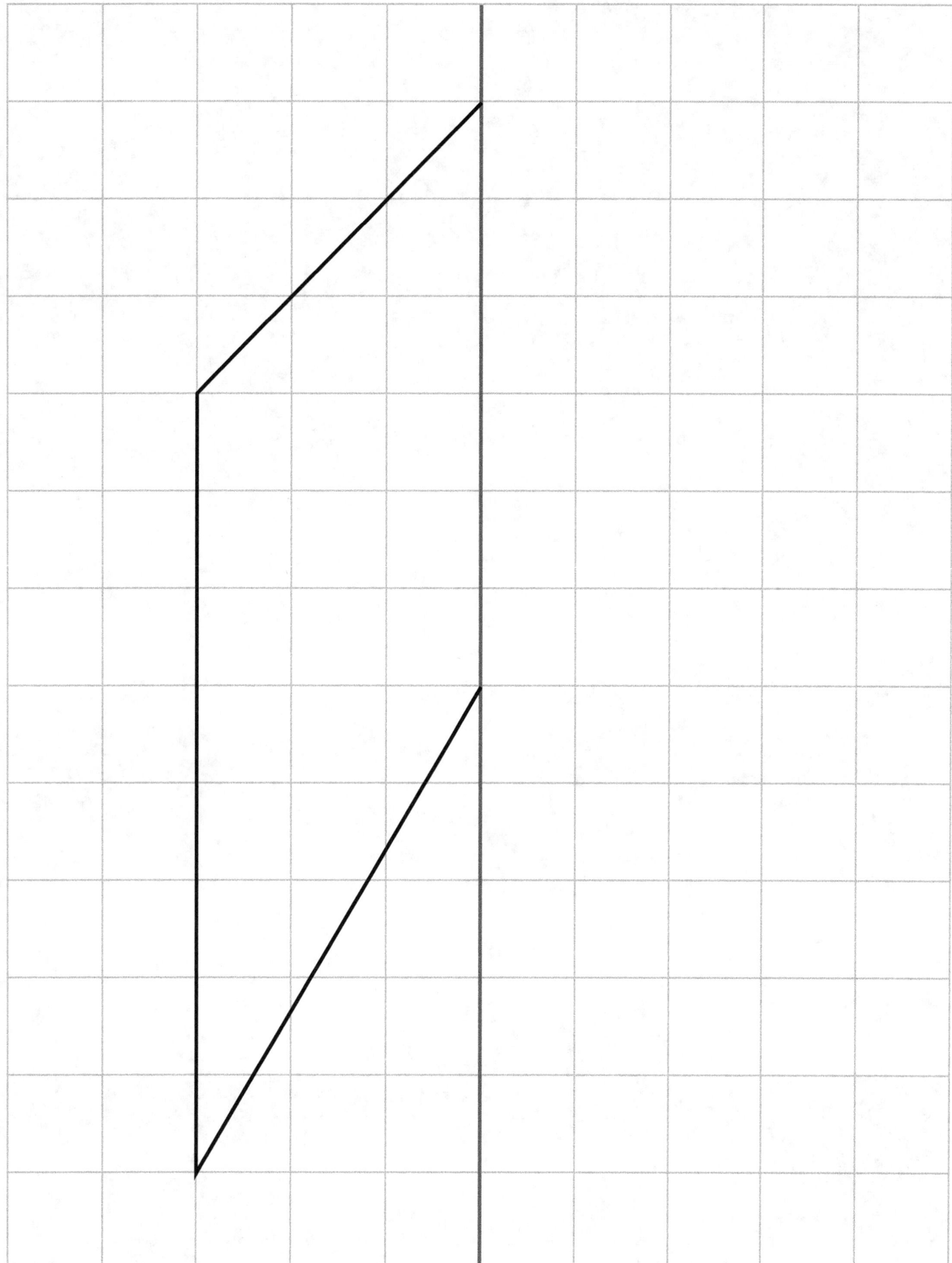

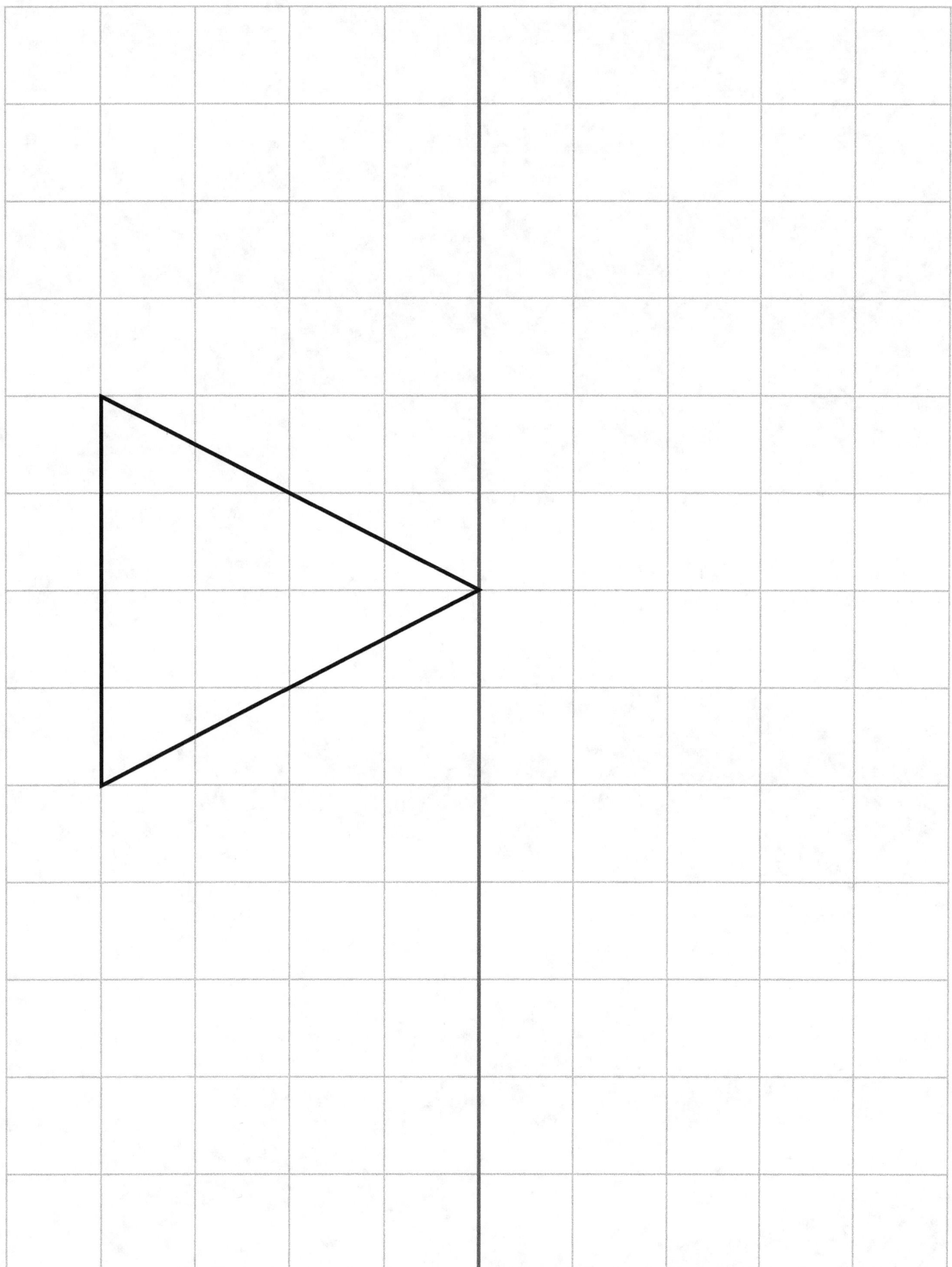

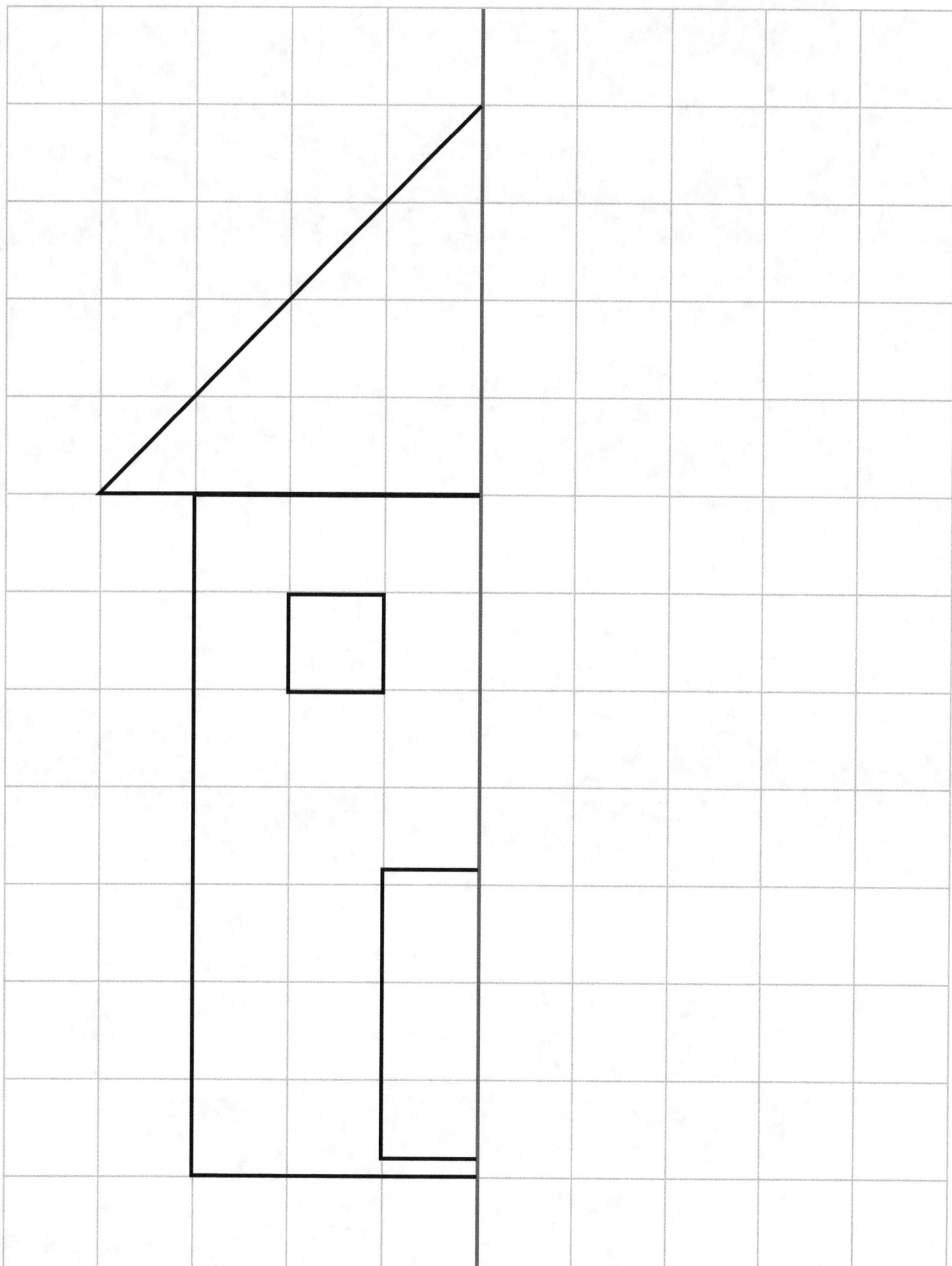

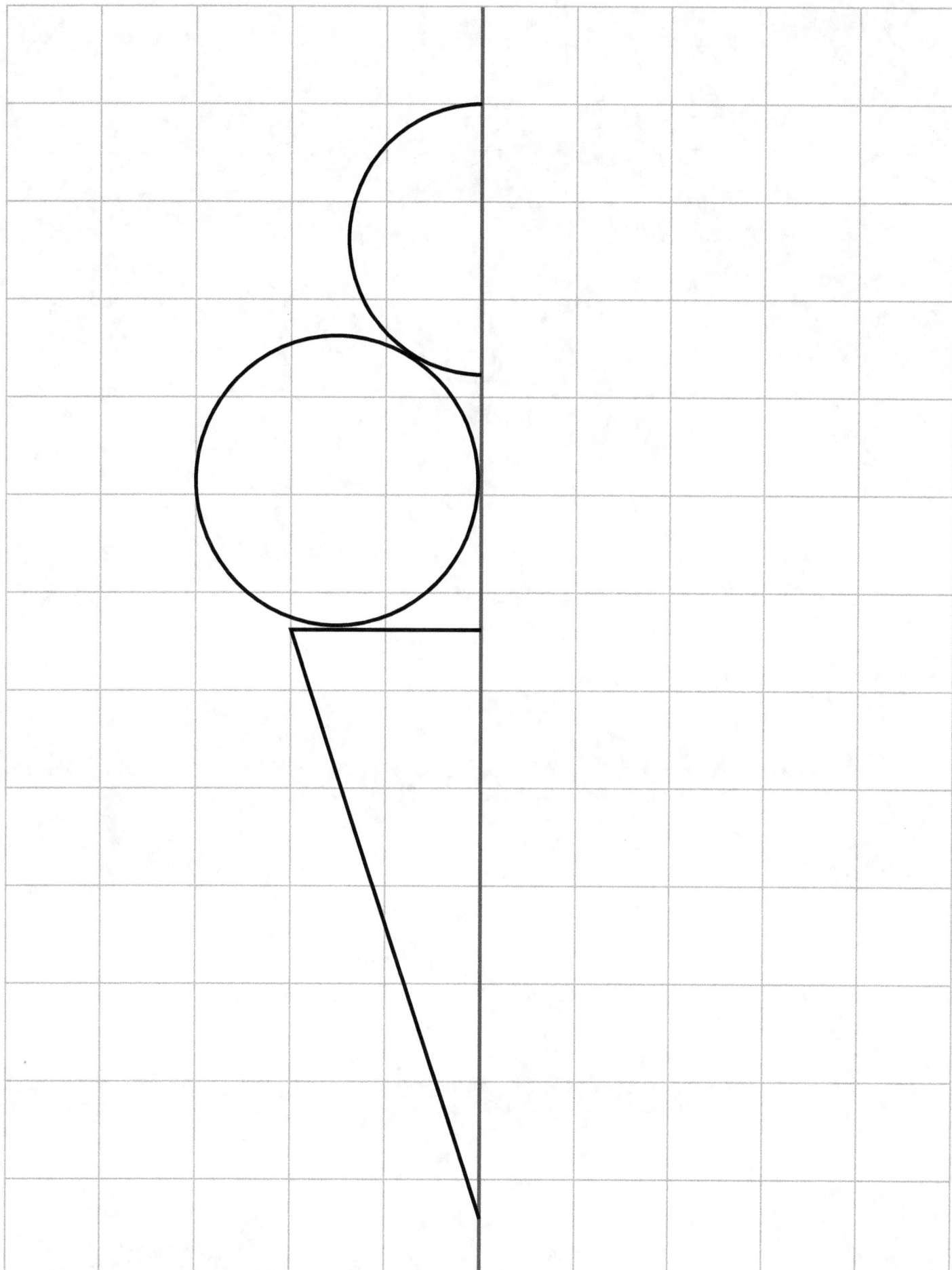

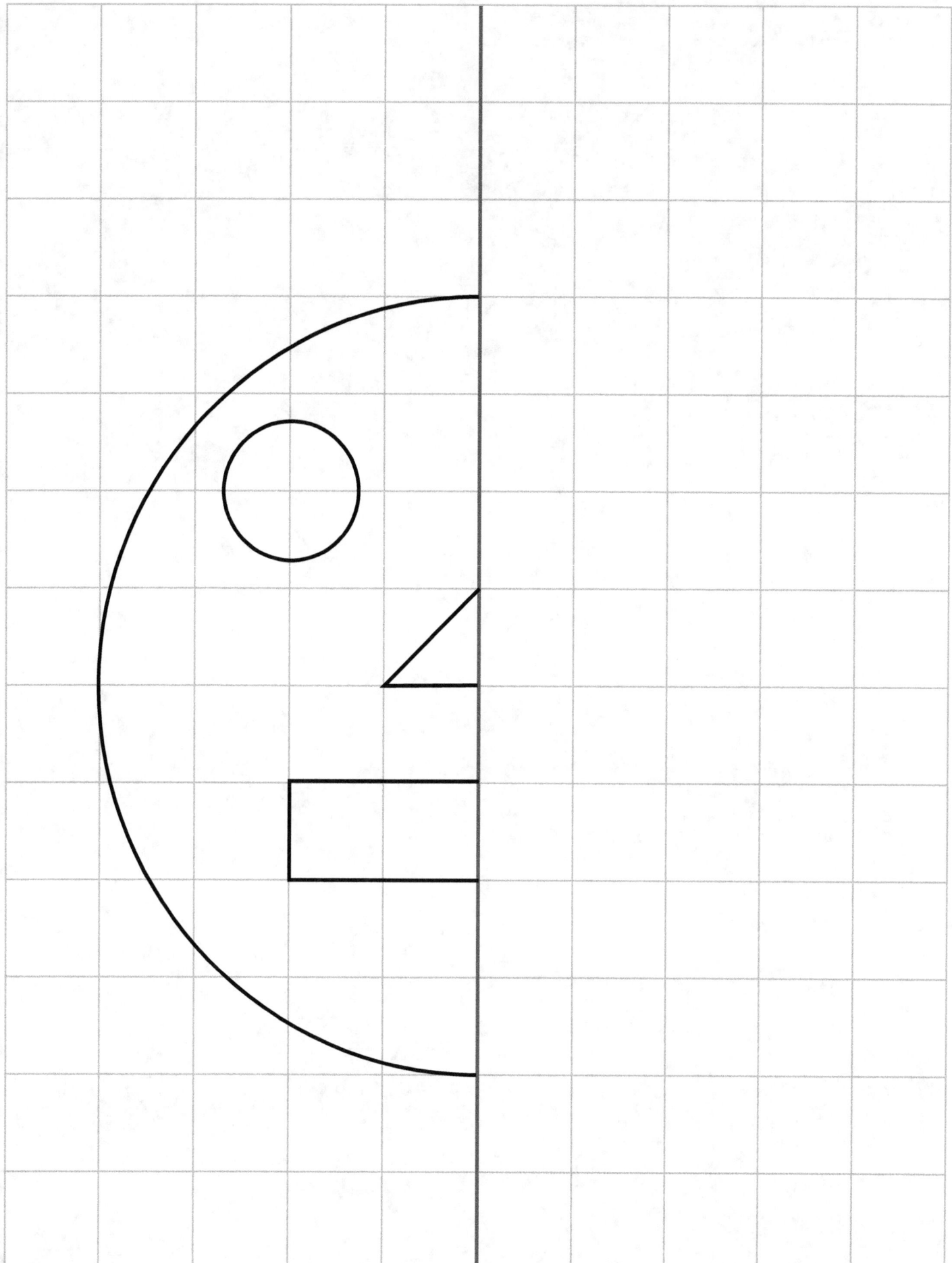

THANK YOU!

Thank you for purchasing this book. We know you could have picked any book, but you picked this one and for that we are extremely grateful.

We hope that it added value and quality to your everyday life. If so, it would be really nice if you could share this book with your friends and family.

If you enjoyed this book and found some benefit in it, we'd like to hear from you and hope that you could take some time to post a review on Amazon. Your feedback and support will help us to greatly improve for future projects and make this book even better.

We'll be more than happy to hear from you, let us know if you have any questions or comments, or tell us to add you to our mailing list for a chance to get **free content, prizes, early sneak peaks and more...**

Just remember, we're always an email away!

info@PinsBasket.com

Wish you all the best in your future success!

Pins Basket team

www.ingramcontent.com/pod-product-compliance
Lightning Source LLC
Chambersburg PA
CBHW081658220526

45466CB00009B/2808